SOME EASY WAYS TO HELP THE ENVIRONMENT!

HANG YOUR CLOTHES OUT TO DRY INSTEAD OF TUMBLE DRYING

FIT ECO-FRIENDLY LONG LASTING LIGHT BULBS

DON'T USE CHEMICALS IN YOUR HOME

USE REUSABLE SHOPPING BAGS

BUY ORGANIC LOCALLY GROWN FRUIT AND VEG

CATCH YOU LATER!

RIDE YOUR BIKE!

EAT LESS MEAT AND DAIRY PRODUCE

BUY SUSTAINABLY CAUGHT FISH

FRESH MILK

BE NICE TO WORMS!

For Margot
You definitely make the world much better!

First published by Scholastic in the UK, 2021
Euston House, 24 Eversholt Street, London, NW1 1DB
Scholastic Ireland, 89E Lagan Road, Dublin Industrial Estate, Glasnevin, Dublin, D11 HP5F

SCHOLASTIC and associated logos are trademarks and/or
registered trademarks of Scholastic Inc.

Text and illustrations © Matt Carr, 2022

ISBN 978 0702 31606 7

A CIP catalogue record for this book is available from the British Library.

Printed in China by C&C Offset Printing

1 3 5 7 9 10 8 6 4 2

www.scholastic.co.uk

FSC
www.fsc.org
MIX
Paper from
responsible sources
FSC® C008047

SCHOLASTIC

DAVE AND GRETA MAKE THE WORLD BETTER!

Matt Carr

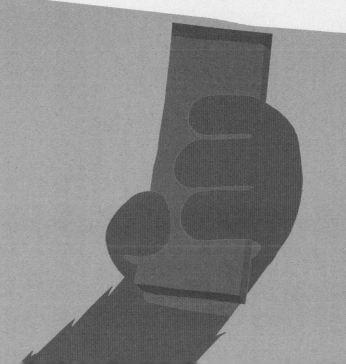

This is the story of **DAVE** and **GRETA**
and how they helped make the world
a little bit better.

Now **DAVE** and **GRETA** were the best of friends
and they lived on the edge of a wood.
Every day was an adventure
and life was pretty good.

Spring, summer, autumn, winter – they had fun no matter the season.

But they noticed that things
weren't quite the same
and Dave soon found out
the reason.

One evening in his underground home, he was watching the news on TV. They were talking about how the weather was different...

...so he called his best friend up the tree.

WALTER WALRUS   20:00

They were both still
rather worried
when they met up
the very next morning.

So they decided to try and think of ways
to keep the world safe from
GLOBAL WARMING.

Dave thought they'd need something **BIG.**
He had lots of great stuff in his head...

...but Greta wanted to find out more
so she went to the library instead.

Dave went to share
his ideas the next day.
He was feeling very proud.

He pinned up his efforts on the notice board...

...and they drew a bit of a crowd!

But instead of being impressed, they all began to laugh.
Nigel the bear was the loudest.

These ideas are really **DAFT!**

I don't know why you're bothering,

Nigel chuckled to Dave.

If the weather gets too hot
or there's too much rain,
We can all just
go back to my cave!

Dave felt very embarrassed.

My silly plans are just pie in the sky.

I'M FAR TOO SMALL TO MAKE THINGS CHANGE

no matter how hard I try!

But suddenly there was a **FLAP, FLAP, FLAP!**

It was Greta
the wise young owl.

DON'T WORRY!
We can still make
the world better.
I'm here to show
you how!

LITTLE things can make a **BIG** difference...

...like saving energy by turning off lights.

Or recycling bottles, cardboard and tins.

This can all help to make things right.

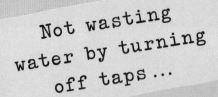

Not wasting water by turning off taps...

...and re-using paper too.

THESE ARE JUST SOME WAYS TO HELP THE ENVIRONMENT AND THERE ARE MANY MORE THINGS YOU CAN DO!

Going **GREEN** is so **EASY!**

laughed Greta.

Even for old bears like you. You can start by not using too much toilet roll every time you go to the loo!

But the most important thing of all,

said Greta, the brave little bird.

Is to care for nature everywhere, and make sure you...

And so Dave and Greta and all their pals
became an ECO-friendly crew
and showed that **EVERYONE** can help the planet
and **YOU** can do it too!

TREE-MENDOUS TREES!

Trees soak up carbon dioxide

TREES ARE AMAZING THINGS!

They make oxygen

They are home to lots of wildlife

They have cool seeds!

OAK: ACORN

HORSE CHESTNUT: CONKER

SYCAMORE: HELICOPTER
(Actually flies like a helicopter!)

PLANT YOUR OWN OAK TREE!

1 Collect some acorns

2 Plant four in a pot of compost

3 The next spring seedlings should grow

4 After two or three years plant out the saplings

IT'S EASY!

DON'T FORGET TO WATER THEM!